YOUR KNOWLEDGE HAS VALUE

- We will publish your bachelor's and
  master's thesis, essays and papers

- Your own eBook and book -
  sold worldwide in all relevant shops

- Earn money with each sale

Upload your text at www.GRIN.com
and publish for free

Thomas Lekscha

Ultrasound for medical therapy devices

Technical Investigation

GRIN Publishing

Imprint:

Copyright © 2010 GRIN Verlag GmbH
Print and binding: Books on Demand GmbH, Norderstedt Germany
ISBN: 978-3-640-74520-3

Ultrasound for medical therapy devices

Technical Investigation

Thomas Lekscha

Introduction

List of Sources

A   Introduction

The ultrasonic therapy appertains to the procedures of the Physical Therapy.
It is a part of mechanical therapy, because ultrasound consists of acoustic waves
(longitudinal, wavelike spreading of smallest pressure variations of a medium as
for example air or liquids) with a high frequency.
In addition the ultrasonic therapy can be used as thermotherapy. The ultrasonic
therapy is also, in the widest sense, a form of the electrotherapy. This justifies
itself, because the ultrasound is being obtained from electric energy.
The therapy form is used particularly with chronically degenerative illnesses of
the human movement apparatus.

The background to this technical summary is based on the legally prescribed
tests on technical medical devices. According to the (MPBetreibV) Medical
Devices Directive[1], ultrasonic therapy devices must be subjected to regular
safety inspections (STK). For these safety inspections, the basics of the technical
ultrasonic therapy should be known.

This abstract shall give a basic overview of the ultrasound and of the transfer
chain; from the ultrasound to the human skin.

---

[1] Directive concerning the setting up, operation and use of medical devices (MPBetreibV)

B   Basics of Ultrasound

Ultrasound relates to material oscillations (periodic, successive pressure fluctuations in conductive media) beyond the upper limit of human hearing. Sound waves above between 20 kHz to 1 GHz are referred to as ultrasonic waves. Distinction of the sound according to the frequency range [1]:

- Infrasound < 16 Hz
  (not audible for humans, too low-frequency).
- Audible sound from16 Hz to 20 kHz
  (audible for humans).
- Ultrasound from 20 kHz to 1 GHz
  (not audible for humans, too high-frequency).
- Hypersound > 1 GHz
  (sound waves with only a limited propagation capability).

B.1   Selected definitions of terms

- Longitudinal wave
  A longitudinal wave is a pressure wave in which the direction of oscillation of the moving particles (molecules) is identical to the direction of propagation. So called pressure fluctuations or density variations are produced within an elastic medium. Sound waves in gases and liquids are always longitudinal waves [2].

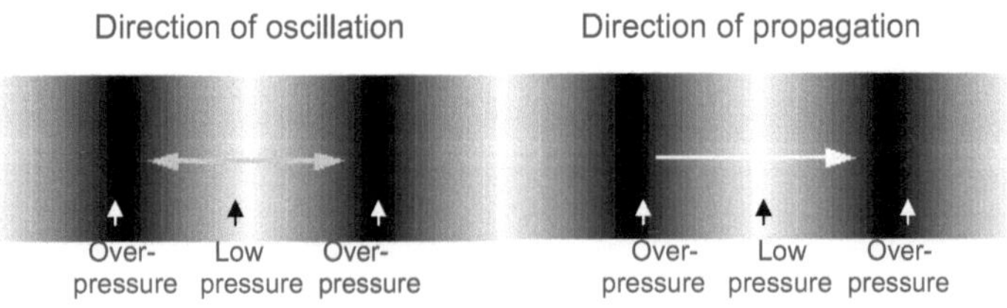

Fig. 01
Illustration of a longitudinal wave. Direction of oscillation and direction of propagation are similar

- Transverse wave

  A transverse wave is a physical wave in which the direction of motion of the
  oscillating particles (molecules) is different from that of the direction of
  propagation. The molecules oscillate perpendicular to the direction of
  propagation.

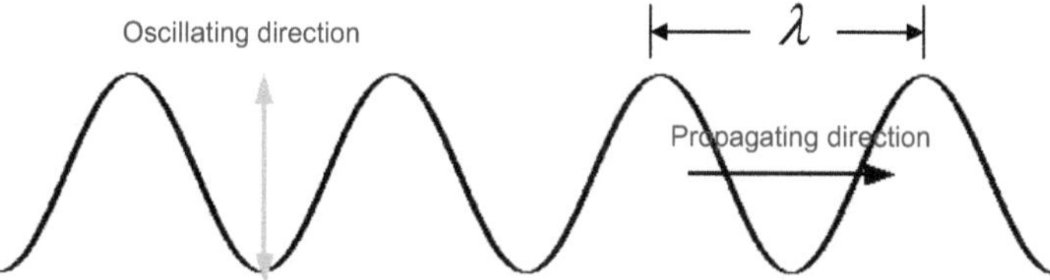

Fig. 02
Illustration of a transverse wave. Direction of oscillation and direction of propagation are
dissimilar

- Sound wavelength

  The wavelength λ is the shortest distance between two points in the same
  phase (identical alignment and direction of movement) of a wave. In the
  example depicted in Fig. 02, the distance between two neighbouring peaks.

$$\lambda = \frac{c}{f}$$

c = sound propagation velocity
f = frequency of the sound wave

(F.1.0)

- Sound frequency

  The frequency of the wave corresponds to the number of waves which pass
  a certain point per unit of time. In the case of oscillation, they are measured
  per second. Within the scope of ultrasonic therapy, the frequency of the
  sound plays a major role because the lower the frequency, the deeper the
  penetration into human tissue (and vice versa) [3].

$$f = \frac{c}{\lambda}$$

c = sound propagation velocity
λ = sound wavelength

(F.1.1)

- Sound velocity

Sound velocity relates to the propagation rate of the sound wave in a specific medium. The sound velocity is determined by the density of the medium being penetrated and its bulk modulus. The bulk modulus, together with the rigidity modulus and torsion modulus, is responsible for the elastic behaviour of a medium. The bulk modulus is a temperature and pressure dependent material constant and can be obtained from mathematical tables. The sound velocity increases with the rigidity of the material or medium (Table 01).

In the case of gases and liquids (no rigidity), the rigidity and torsion modulus can be set equal to zero and, as a result, no transverse waves can propagate but only longitudinal waves.

$$c = \sqrt{\frac{K}{\rho}}$$

K = bulk modulus of the medium

ρ = density of the medium

(F.1.2)

- Acoustic impedance

Acoustic impedance or characteristic acoustic impedance is also referred to as the resistance of a specific material to a sound wave. The characteristic acoustic impedance is the product of the density of the medium to be penetrated and the specific sound velocity.

$$Z = \rho \cdot c$$

ρ = density of the medium

c = sound velocity (in the medium)

(F.1.3)

Material		Density kg/m^3	Sound velocity m/s	Impedance Ns/m^3
Air	(20°C)	1.21	344	416
Air	(40°C)	1.13	387	436
Water	(20°C)	998	1483	1.47 x 10^6
Water	(40°C)	992	1529	1.52 x 10^6
Fat tissue		920	1410-1479	1.33 x 10^6
Muscle tissue		1060	1540-1603	1.67 x 10^6
Bone		1380-1810	2700-4100	4.3-6.6 x 10^6
Steel		7850	6000	47.1 x 10^6

Table 01
Summary of the density, sound velocity and acoustic impedance of media [4,14]

- Sound pressure

  Sound produces longitudinal <u>pressure</u> waves in liquids and gases. The pressure wave exerts a certain force on a defined surface. An ultrasonic therapy probe sends a sound wave on a defined area of tissue. In this case, the ultrasonic wave (pressure fluctuations) exerts a specific force per area of tissue. This quotient is referred to as sound pressure.

$$p = \frac{F}{A}$$

F = force

A = area

(F.1.4)

- Ultrasonic sensor

  An ultrasonic sensor is a component capable of detecting changes in physical magnitudes (e.g. pressure fluctuations). The sensor converts these changes in physical magnitudes to electrical signals and sends them to the relevant evaluation unit. Refer to Fig. 08 on Page 12.

- Sound actuator

  A sound actuator is the counterpart of the sound sensor. A sound actuator is a component which converts electronic signals to a mechanical action. Refer to Fig .09 on Page 13.

- Coupling

  Coupling relates to the adaptation of various impedances or characteristic acoustic impedances to one another. Ultrasonics, here, is analogous to electrotechnics (internal resistance = load resistance = optimum impedance adaptation). A coupling medium is used (water or ultrasound gel) in an attempt to eliminate, or reduce as far as possible, transition resistance from the ultrasonic transducer to the tissue. If the acoustic impedance of the ultrasonic transducer was practically the same as the acoustic impedance of the tissue, the sound could be transferred to 100% in an ideal case.

A reflection of the ultrasound against bordering surfaces could, thus, be prevented. However, a 100% adaptation is not possible in practical applications.

C   Basics of Ultrasonic Therapy

In order to implement ultrasonics for medical treatment, an ultrasonic therapy device is required. A wide range of such therapy devices is currently available on the international market. However, the structure of the basic modules used in these devices is always the same. In order to produce an ultrasound, one needs a high-frequency generator and an ultrasonic transducer with an integrated oscillating crystal (piezoelectric actuator). The high-frequency generator produces an alternating voltage which is transferred to the oscillating crystal. This alternating voltage causes the oscillating crystal to change its geometrical shape (due to the piezoelectric effect). The oscillating crystal oscillates, and sends sound waves in the ultrasonic range. To achieve a broader range of application possibilities, the oscillating crystal can be excited either continuously or pulsed. The corresponding operating mode can then be selected on the ultrasonic therapy device according to the indicators. It is worth observing, however, that the human tissue warms up more when continuous ultrasound is transmitted than in the case of pulsed sound transmission.

The output frequency of most therapy devices is restricted to 1 MHz and 3 MHz. These two frequencies and their respective modulations are sufficient to treat numerous medical conditions. Normally, each device is equipped with a separate ultrasonic button for each frequency.

In order to be able to emit the sound waves on the human body, a coupling medium is required for the transfer (coupling the ultrasonic transducer to the human tissue). In practice, ultrasound gel or water is generally used as the ultrasound coupling medium. Since the various media (ultrasonic transducer, air, tissue) conducts and absorbs the ultrasound differently, selection of the right coupling medium is of great importance.

C.1 Effects on and in the human tissue

Ultrasonic therapy produces different effects on the human body and in human tissue. In general, a distinction is made between two effects:

- Mechanical effects
  The ultrasound output or ultrasonic waves applied to human tissue produce contractions and expansions in the elastic human tissue using the same frequency (1 MHz or 3 MHz). This effect is also referred to as a micro-massage. The micromassage causes a volume change in cells, increases cell membrane penetrability and, thus, an increase in the exchange of metabolic products. Blood circulation is aided and accelerated by a micro-massage, thus achieving better "sustenance" of the tissue.

- Thermal effects
  The thermal effects (warming of body tissue) are a result of the various characteristics (density, fat and water content) of the tissue types. The temperature increase is approximately proportional to the density and specific capacity of the tissue.
  Together with the absorption coefficient α [2], the sound energy radiated into the tissue is the decisive parameter with regard to the level of warming of the tissue [5]. The greatest thermal effect occurs at the boundary layer of various tissue structures due their differing impedance (resistance exerted on the sound wave). The higher the absorption coefficient, the greater the warming effect in the tissue.

---

[2] α specifies the extent to which a physical magnitude is absorbed in a damping medium

Medium	Absorption coefficient at 1 MHz	Absorption coefficient at 3 MHz
Bone tissue	3.22 cm^{-1}	-
Skin	0.62 cm^{-1}	1.86 cm^{-1}
Water (20°C)	0.0006 cm^{-1}	0.0018 cm^{-1}
Air (20°C)	2.76 cm^{-1}	8.28 cm^{-1}
Muscle tissue	0.76 cm^{-1}	2.28 cm^{-1}

Table 02
Absorption coefficients depending on medium and frequency [3]

Medium	Absorption coefficient at 1 MHz	Absorption coefficient at 3 MHz
Cartilage	1.16 cm^{-1}	3.48 cm^{-1}
Tendon	1.12 cm^{-1}	3.36 cm^{-1}
Fat	0.14 cm^{-1}	0.42 cm^{-1}
Nerve	0.2 cm^{-1}	0.6 cm^{-1}

Table 03
Absorption coefficients depending on medium and frequency [6]

The third effect of ultrasonic therapy should be mentioned at this point, the so called "physiological" effect.

This effect results from the two above effects and cannot be unequivocally defined or assigned. Even the two interdependent effects described above in conjunction with human tissue cannot always be observed and treated separately. Usually, the interaction of the two mechanisms plays a decisive role in the therapeutic effects of ultrasound treatment.

The application of ultrasound in treating the human body is mainly described by its frequency, modulation curve and output power.

Information on indications and treatment parameters or treatment period is generally provided in the respective device operating manuals enclosed with the various ultrasonic therapy devices. They also recommend device-specific settings for the therapeutic treatment indicated.

C.2   The effect spectrum and indications

Ultrasonic therapy is a technical, medical therapeutic procedure which is usually implemented for heat and stimulation applications locally or regionally. Due to the effects described above (Pages 7 to 8), a whole range of positive, healing effects can be provided to the human body.

Effects:
- Pain relief
- Stimulates blood circulation
- Releasing tissue adhesions
- Warming tissue areas
- Relaxing muscles
- Supporting healing of fractures. [7]

The above list only reflects a selection of effects on human tissue and does not claim to represent a complete list.

An ultrasonic therapy application can be considered appropriate for the following patterns and courses of illnesses.

Indications [3] :
- Muscle tension
- Arthrosis of the knee and hip
- Scar tissue adhesion
- Inflammation and strain of tendons
- Tennis and/or golfer's elbow. [8]

The above list only reflects a selection of frequent indicators and does not claim to represent a complete list.

---

[3] Application of a specific medical measure for a specific cluster of symtoms

Pictures of example applications of ultrasonic therapy:

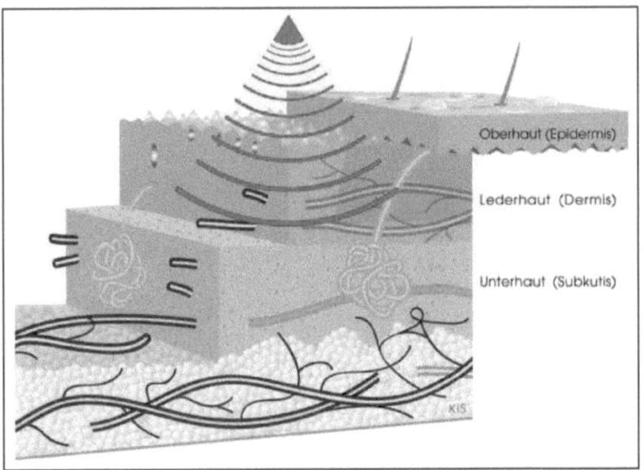

Fig. 03
Application of ultrasound on human tissue with illustration of the layers of skin

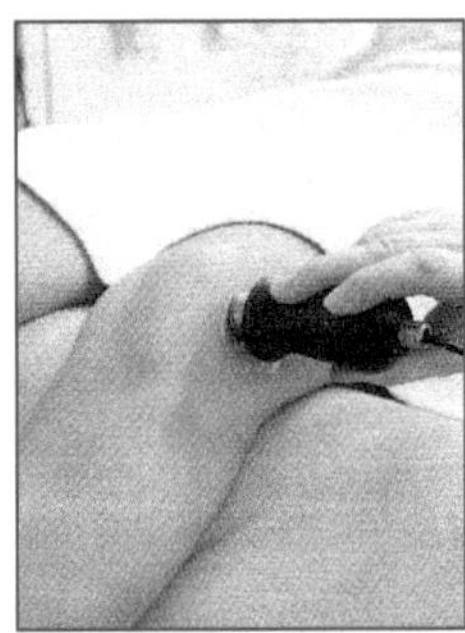

Fig. 04
The picture shows the application of ultrasonic therapy
on a knee to treat gonarthrosis

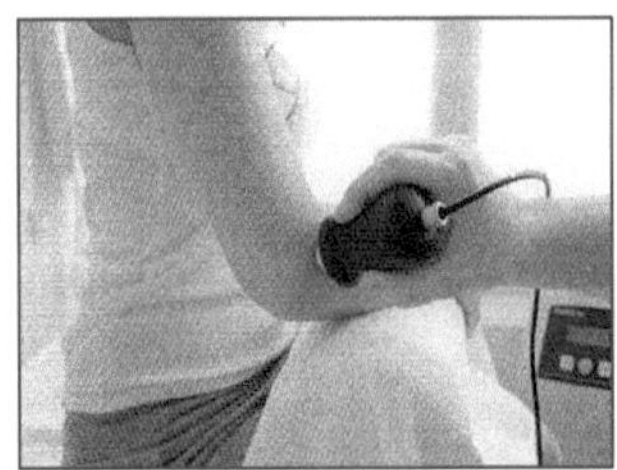

Fig. 05
The picture shows the application of ultrasonic therapy
on an elbow to treat tennis or golfer's elbow

D   Basics of the Piezoceramic Sensor

D.1   Historical excursion

The piezoelectric effect was discovered by the French physicists Pierre and Jacques Curie in 1880. The two brothers experimented with natural crystals (salt, tourmaline, quartz) and discovered that when certain crystals are mechanically deformed, they are electrically charged or polarised.

They also found that by applying an electric current to a crystal, it deformed. This reciprocal effect was described as the inverse piezoelectric effect.

The discoveries of the French physicists formed the basis for the sonar, developed in around 1940 for navy vessels to locate submarines.

The breakthrough in piezotechnology was made by Russian and American scientists in the years leading up to the 1950s as they succeeded synthesising the first PZT (plumbum-zirconate-titanate) compounds.

These compounds remain the dominating materials for piezotechnology thanks to their excellent properties [9].

The discoverers of the piezoelectric effect

Fig. 06
French physicist
Jacques Currie (1855-1941)

Fig. 07
French physicist
Pierre Currie (1859-1906)

D.2   Principles of functioning

Oscillating crystals (piezoceramic components) are used to produce ultrasound for medical applications. Oscillating crystals can be made of various materials, such as quartz, barium titanate and lead-zirconate-titanate. When piezoelectric materials are mechanically deformed by an external force, electric charges are produced on their surfaces due to the change of polarization. The greater the deformation or application of force, the greater the voltage which can be tapped. The generation of electrical charges related to deformation is referred to as the direct piezoelectric effect.

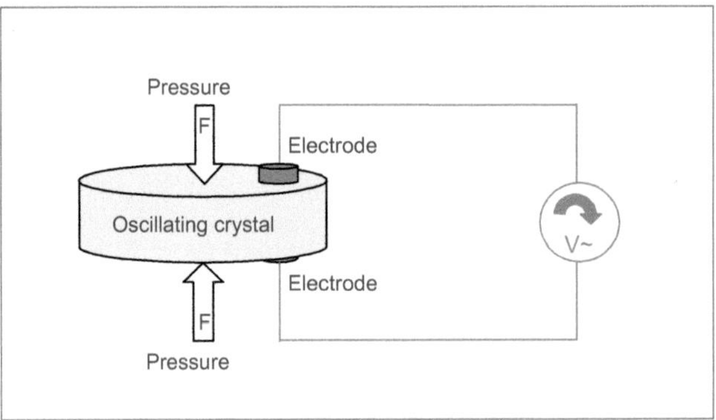

Fig. 08
Illustration of an oscillating crystal (piezoelectric disc) on which a mechanical force is exerted. The mechanical deformation energy is converted to electrical energy (separation of charge) which can be measured

This effect is also reversible. If, for example, an alternating voltage is applied to an oscillating crystal, the geometrical shape of the material changes as a result (stretching, compression). These changes of shape can cause pressure or sound waves in the ultrasonic range, whereby the alternating pressure is dependent on the alternating voltage applied, the frequency of the alternating voltage and the geometrical force of the oscillator. This process is referred to as the inverse piezoelectric effect.

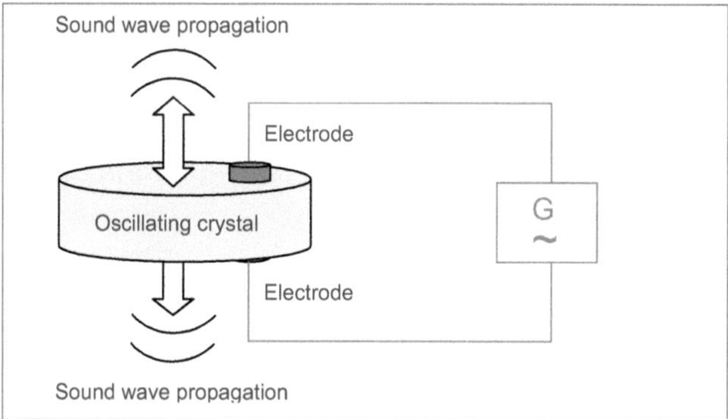

Fig. 09
Illustration of an oscillating crystal (piezoelectric disc) and its alternating alignment (pressure waves) when an alternating voltage is applied

D.3   Production and material

The production of piezoceramic sensors is heavily dependent on the use for which the sensors are intended. However, the typical sequence of processes for the production of piezoceramics is similar for all types. Production always begins with the selection and mixture of the basic materials (normally on a lead-zirconate-titanate basis). The mixture is then heated to approx. 800°C to 900 °C. This heating or firing process is also referred to as calcination. After calcination, the mixture, made up of various ingredients, is ground. This converts the mixture into fine, powdery particles, necessary for subsequent processing.
The powder is then pressed to produce the rough form required. Since the blanks cannot yet be exposed to any mechanical loads, they must be "baked" at approx. 1000°C to 1300°C. This process is also referred to as sintering. The polycrystalline ceramic structure is formed during the course of sintering. After sintering, the ceramics are set to their final form by means of a mechanical process. During this working process, the ceramics are sawn, ground and polished. When the ceramics have been brought to their final shape, the final working process is performed, whereby the crystals are polarized. During polarization, the dipole in the material is aligned by applying a direct current electrical field of approx. 2000 V to 3000 V.

This alignment of the dipoles is retained (almost completely) even after switching off the direct current electrical field. Polarization occurs at temperatures between 80°C and 140 °C.

The piezocrystal has now been produced [10].

The final step involves measuring the crystal and assigned to certain accuracy classes according to its tolerances or, in the case of failure to comply with the tolerances, the crystal is rejected.

General summary of the production steps to produce a piezoceramic oscillator:

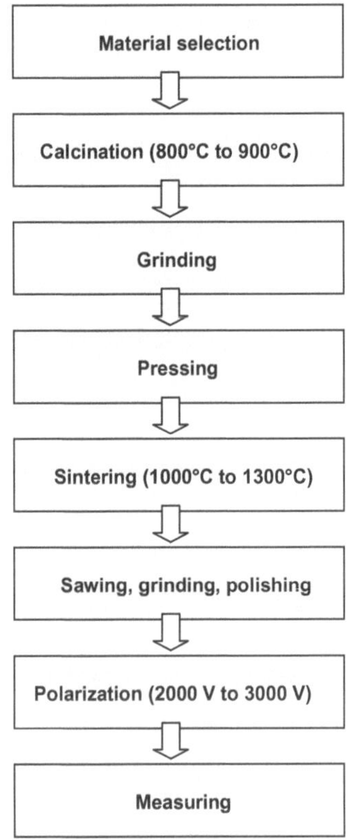

Fig. 10
General summary of a typical production process to produce piezoceramic crystals

Today, piezoceramic sensors for use medical applications as well as high-performance ceramics are almost exclusively produced from the lead-zirconate-titanate compound. This compound is made up of lead (Pb), oxygen (O), zirconium (Zr) and titan (Ti). This compound is often referred to by the abbreviation of its parts, namely "PZT". The advantage of this compound or mixed crystal is that it is electrically neutral above the Curie point [4]. Below the Curie point, the titan atom moves from its central position and the previously neutral grid becomes a dipole. This dipole grid has piezoelectric characteristics and can be used to provide excellent, low-cost actuators in medical technology [11].

Piezoceramic sensors can also be produced implementing materials with special characteristics (heat resistant, unbreakable, flexible, etc.) for use in very specific applications. However, this study is not concerned with these special production processes or special mixed crystals. The PZT-materials described above are the mixed crystals most frequently used in medical device technology.

D.4   Forms used and their basic oscillation

Oscillating crystals can be produced in various forms, from different materials and with preferential propagation directions.

Selection of the correct form and material relates to the respective task at hand. The deflection or deformation direction of the mechanical deformation is compelled by the direction of polarization.

Discs, punched discs, rectangular plates, bars, cylinders and pipes represent the typical geometric forms for piezoceramic oscillators.

Some frequently used geometric forms and their deformation directions are illustrated in the figure on Page 16.

---

[4]   Temperature at which the relative dielectric constant of ferroelectric ceramics reaches its maximum. At this temperature, piezoelectric ceramics lose their polarization condition applied during production [12]

Basic oscillating forms of piezoelectric oscillators

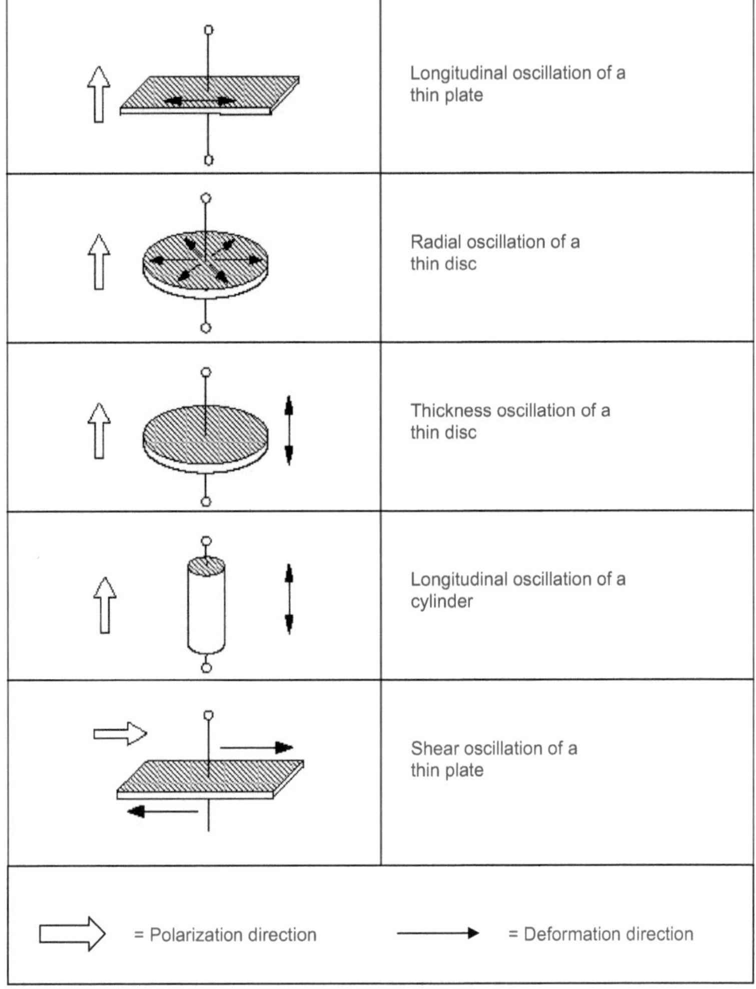

Fig. 11
Comparison of typical, basic geometrical forms and their deformation directions
related to piezoelectric oscillators [13]

Ultrasonic transducer for medical technology

The following pictures show a typical medical transducer for ultrasonic therapy.

Fig 12
The ultrasonic therapy transducer
side view

Fig. 13
The ultrasonic therapy transducer
plan view

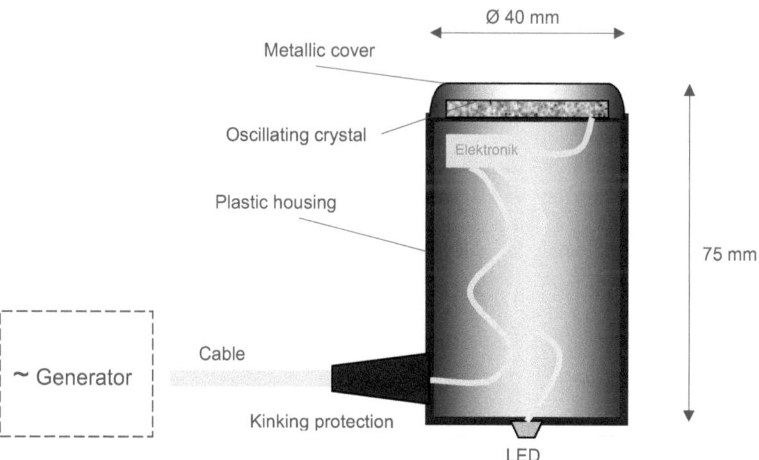

Fig. 14
Illustration of the main components of an ultrasonic therapy transducer

F   List of Abbreviations

Abbreviation	Explanation	Comment
MPBetreibV	Medical Devices Directive	German
STK	Safety-related controls	German
approx.	approximate(ly)	
e.g.	exempli gratia	
PZT	Plumbum-zirconate-titanate	
Pb	plumbum = lead	Chemical element
O	oxygenium = oxygen	Chemical element
Zr	zirconium	Chemical element
Ti	titan	Chemical element
LED	Light Emitting Diode	
Fig.	Figure	Fig. 01 to Fig. 14
F.1…	Formulas	F.1.0 to F.1.4

G   List of Equation and Formula Symbols / Units

Quantity		Formula Symbol	Units	Relation Conversion
Wavelength	(F.1.0)	λ	m	10^3 mm 10^6 µm
Sound Propagation speed	(F.1.2)	c	m / s	3.6 km / h
Frequency	(F.1.1)	f	Hz	1 / s
Compression module	(F.1.2)	K	Pa	N / m^2
Density	(F.1.2)	ρ	kg / m^3	10^{-3} g / cm^3 10^3 g / m^3
Characteristic impedance	(F.1.3)	Z	N s / m^3	kg / (m^2 s)
Sound pressure	(F.1.4)	p	Pa	N / m^2
Force	(F.1.4)	F	N	kg m / s^2
Area	(F.1.4)	A	m^2	10^6 mm^2 10^4 cm^2
Absorption coefficient		α	m^{-1}	1 / 10^3 mm

H   List of Tables

I   List of Figures

J   Bibliography

[1]   Wikipedia ® Wikimedia Foundation Inc.:
                Schall, Einteilung nach Frequenz
                Onlinepublikation: http://de.wikipedia.org/wiki/schall
                Abgerufen: 17.02.2010

[2]   Harten,U.: Physik für Mediziner. 12. Aufl., Seite 112,
                Berlin, Heidelberg: Springer 2007, ISSN 0937-7433

[3]   Steiner, Schneider: Physikalische und technische
                Grundlagen der Dopplersonographie. 2. Aufl., Seite 13
                Berlin, Heidelberg: Springer 2008, ISBN 978-3-540-72370-7

[4]   Hering, Stohrer: Physik für Ingenieure. 10. Aufl., Seite 617,
                Berlin, Heidelberg: Springer 2007, ISBN 978-3-540-71855-0

[5]   Huber, Debus, Jenne: Therapeutischer Ultraschall in der
                Tumortherapie. Grundlagen, Anwendungen und neue
                Entwicklungen. Heft 36 DOI 10.1007/S001170050041
                Berlin, Heidelberg: Springer Januar 1996,

[6]   Onlinepublikation: www.physiosupport.org/us.htm
                Abgerufen: 17.02.2010

[7,8]   Revitalis, Ultraschalltherapie
                Onlinepublikation: www.revitalis.de/physiotherapie/
                ultraschalltherapie.php. Abgerufen 17.02.2010
                Lippstadt

[9]   PI Ceramic GmbH: Tutorium Piezoaktorik und
                Nanopositionierung, Germany

[10,11,12]   CeramTec AG, Geschäftsbereich Multifunktionskeramik:
                Hochleistungskeramik in der Piezotechnik
                Printed in Germany: MF080007.D/GB.2000.0807

[13]   DIN EN 50324-1: Dezember 2002
                Piezoelektrische Eigenschaften von keramischen
                Werkstoffen und Komponenten
                DIN Deutsches Institut für Normung e.V., Berlin

[14]   Hering, Modler: Grundwissen des Ingenieurs. 13. Aufl.
                Leipzig: Fachbuchverlag, ISBN 3-446-21443-7

[15]   DIN EN 50324-1: Dezember 2002
                Piezoelektrische Eigenschaften von keramischen
                Werkstoffen und Komponenten
                DIN Deutsches Institut für Normung e.V., Berlin

[16]   Frey, R. Onlinepublikation: http://commons.wikimedia.org/wiki/
         File:Longitudinalwelle_Transversalwelle.png
         Bern. Abgerufen 05.01.2010

[17]   Zwehn,S., Onlinepublikation:http://sabinezwehn-cosmetics.de/
         attachments/Image/Haut-Grafik-Ultraschall.jpg
         Bingen-Büdesheim. Abgerufen 05.01.2010

[18]   Bandelin electronic: Schmerzen „wegschallen" mit der
         Ultraschall-Therapie.
         Berlin. Flyer-Drucknummer: 64920/2006-09

[19]   Tel Aviv University: Piezoelectric History. Onlinepublikation:
         http://www.tau.ac.il/~phchlab/experiments/QCM/Piezoelectric_
         History.htm. Abgerufen 10.01.2010